TERMITES AND BORERS

PHIL HADLINGTON AOM is a consultant in pest control and tree care. He retired from a distinguished career in research as a forest entomologist in New South Wales, having also played the most significant and influential role in educating not only the pest management industry but ordinary homeowners who attended his many courses and read his many books. He was awarded the Order of Australia Medal in 2000 for his long-term contribution to the Australian community directly, and indirectly, through his education of the pest management industry which in turn serves the community.

ION STAUNTON was a pest management technician before he became the founding secretary of the NSW and national associations. His management and organisational skills helped to overcome the competing-company wariness of the 1960s so that the industry is now focused on the goal of providing the best possible service for its customers. Ion recently invented a windowed termite monitor and treatment station that, for the first time, allows homeowners to be vitally involved in reducing the threat from nests which surround their homes.

PHIL HADLINGTON and ION STAUNTON produced the first industry text book in 1960.

TERMITES AND BORERS

A HOMEOWNER'S GUIDE TO DETECTION AND CONTROL

PHILLIP **HADLINGTON**
ION **STAUNTON**

UNSW PRESS

CONTENTS

7
Acknowledgements

9
Foreword *by*
Doug Howick

11
Introduction

13
1 – TERMITE BIOLOGY
AND HABITS
Termite castes
Behavioural patterns
Termite pest species
Nesting sites

32
2 – PROTECTING
YOUR HOUSE
Design
Physical barriers
Chemical barriers
Monitors and
      treatment devices

A UNSW PRESS BOOK

Published by
University of New South Wales Press Ltd
University of New South Wales
Sydney NSW 2052
AUSTRALIA
www.unswpress.com.au

© Phillip Hadlington and Ion Staunton 2006
First edition 1998
Reprinted 1998 and 2002
This edition published 2006

This book is copyright. Apart from any fair dealing for the purpose of private study, research, criticism or review, as permitted under the Copyright Act, no part may be reproduced by any process without written permission. Inquiries should be addressed to the publisher.

National Library of Australia
Cataloguing-in-Publication entry

Hadlington, Phillip W., 1923– .
Termites and borers: a homeowner's guide
to detection and control.
2nd ed. Includes index.
ISBN 0 86840 827 1.

1. Termites - Australia. 2. Termites - Control - Australia. 3. Borers (Insects) - Australia. 4. Borers (Insects) - Control - Australia. 5. Buildings - Pest control - Australia. I. Staunton, Ion. II. Title.

628.96570994

Design Di Quick
Printer Everbest

CONTENTS

45
3 – INSPECTING FOR TERMITES
What homeowners can do
What the professionals can do
Exclusions
Three different professional inspections
Repairs and undiscovered damage

54
4 – WHAT TO DO IF TERMITES ARE FOUND
The homeowner
Treatment options
Comparing proposals

65
5 – FAQs ON TERMITES

69
6 – BORERS
Pinhole borers
Powderpost beetles
Furniture beetles
European house borers

76
7 – WOOD DECAY AND DEFIBRATION

79
INDEX

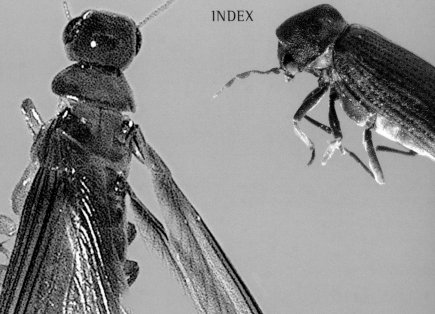

The Australian dream of a home among the gum trees also means a home among the termites.

ACKNOWLEDGEMENTS

Like most professions, timber pest management is continually evolving. Bill Flick is credited as being the first to use arsenic to kill termite colonies in 1915. Before that, if termites got into a house, the owners would do their best to patch up the damage, then move out leaving the problem to the next owner. The Flickmen were well trained in termite control in particular and many of them left Flick and Co with the knowledge to set up their own businesses; hence the industry grew. In the mid 1950s Phil Hadlington, a Forest Entomologist with the NSW state government, began teaching the first pest control course in Sydney. A couple of years later I did that course, and Phil was impressed enough with my drawings to ask me to help with the illustrations for the first industry text book. We then co-authored a pest control correspondence course, which reached technicians in the rest of Australia. Next Phil saw a need for a tree care and tree surgery course, so we did that together and followed up with *Trees for Australian Gardens*, which included a sizeable section on tree surgery. Forty-something years later we team up again.

Our involvement in the education side of the industry has introduced us to many wonderful, knowledgeable and like-minded people who have been very willing to share experiences, research and information. In revising this edition, I called on information and explanations picked up through the years, from my original course to recent articles in *Professional Pest Manager* magazine, visiting stands and attending sessions at many pest management conventions. There are too many names from the past 50 years to list here; so many people, and all so willing to share.

Since the long-lasting organochlorines disappeared into history, shorter-life chemicals have required new approaches to application. New residual membranes and monitoring devices have been developed. To help me catch up with the latest, I acknowledge and sincerely appreciate the willingness of the management of Amalgamated Pest Control (APC) in giving me access to Shaun Hale of their Technical Division. Shaun's suggestions have been a major influence on this book's helpfulness to homeowners. Many APC photographs are used to portray examples of the topics covered. Thank you.

Ion Staunton

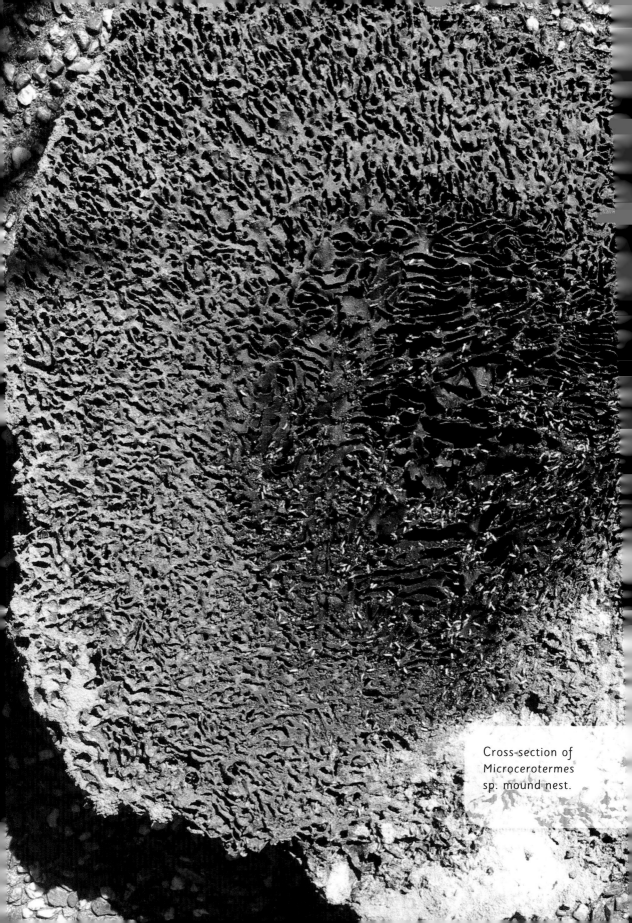

Cross-section of Microcerotermes sp. mound nest.

FOREWORD

I was privileged to be asked to write the Foreword to this book when it was first presented by Phil Hadlington and Christine Marsden in 1998.

Since that time, the technology of termite treatment has improved and expanded, resulting in a wider range of management systems becoming available to Australian homeowners. The biology and habits of termites have not changed. What has changed is the impact of their co-existence with us and their ability to access our most expensive and treasured possessions – our homes.

The knowledge and experience of Phil Hadlington has now been supplemented by that of Ion Staunton. The result is a fresh and refreshing approach to a continuing problem for homeowners. Rather than being a do-it-yourself pest control instruction manual, it fills an important gap by providing sufficient information, in simple terms, to help homeowners understand the potential hazards of timber pest infestation. That knowledge will enable homeowners to make informed judgments about which solutions are appropriate. As the authors declare, even if you find termite activity and break open infested timber or active leads, don't feel good about it – the nest will survive and what's more, they still know where you live!

I believe that this publication will assist discussions between professional pest managers and their clients – and most professional pest managers in Australia are members of the AEPMA.

Doug Howick
National Executive Director
Australian Environmental Pest Managers Association

You rarely see termites, because they shun the light and work in a sealed environment ... unless accidentally discovered.

INTRODUCTION

Very few words so completely and instantly grab the attention of a householder as when a pest management technician or a spouse announces: 'We have termites.'

You have good reason to feel apprehensive. Termites are sneaky. They can infiltrate and secretly undermine the strength and stability of your house. And it's not just the costs of repair; it's the loss of resale value that really hurts your pocket.

Termites are a common problem. A CSIRO survey in 2004 reported that almost 33 per cent of Australian homes had a history of termite attack. The annual damage bill was calculated at $780 million ... 130 000 homes at an average of $6000 each, made up of $1500 for the termite treatment and $4500 for the timber repairs and replacement.

Almost everyone knows a termite horror story. If it's not their own story it's that of a friend or neighbour, or from a current affairs program.

Termites were understandably a big shock to the British settlers of 1788. After more than 200 years, you'd think we'd have worked it out. But we've spent almost all our effort in trying to make our buildings 'termite proof'; making them into fortresses so that the termites can't get in to the timber we use as flooring, frames, fittings and fixtures. The damage bills and horror stories tell us emphatically that the fortress approach fails too often.

This book will explain principles and options to help you minimise failure to your 'fortress' and how the new tools for termite aggregation can simply kill off termite colonies that surround and threaten your house and other structures. You'll learn what to expect from your pest management company, what you can do yourself, and what to look for as you check around.

You will also find helpful information about some borers which attack timber, including treatment options and tips on how to identify them. Wood decay is essentially a water or moisture management problem, and once understood can be eliminated from your worry list. Defibration is not a significant national problem but it could well be happening at your place, so we provide some information on this to round off the book.

TERMITE
biology and habits

CHAPTER 1

'Termite' is the correct and most commonly used word to describe what once were referred to as 'white ants'. Termites are not ants, and only the young nymphs and workers could be called white.

> Ants have elbowed antennae, compared to the termite's 'string of beads'. Termites do not have a 'waist' or constriction behind the thorax.

Termites and ants do have similarities in habit: they are both communal insects; they mostly live in the ground but will invade our structures; and biologically, they each have caste systems with specific functions to ensure the success and survival of their colonies.

Ants are enemies of termites, however the old belief that if you have a lot of garden ants you won't get termites is a myth. Termites construct their nest and working galleries with a dense mud material that ants can't breach, except perhaps in the first few days after the colonising flight when the male and female reproductives are desperately trying to establish and secure their chosen site before ants, birds, lizards, frogs and spiders find them.

Colonising flights of thousands of the termite reproductives leave only from mature nests, five years or more after establishment. The flight is timed for when outside temperature and humidity are most similar to the conditions within the nest. It usually happens in early summer in the evening, after most birds have gone to roost. The flight slits or cuts are usually constructed at a high part of the colony's workings, fairly close to the nest. Soldiers guard these openings by filling the gaps with their heads and jaws until it is time for the reproductives to fly. The slits are resealed immediately after the evening's flight is completed.

Alate termites are the reproductives, ready for the colonising flight.

Termites are weak flyers. Some have a brief attraction to bright lights but as soon as they reach ground level and pair off they shed their wings. Then the search begins for a suitable nesting site. Their criteria, apart from avoiding the predators, are simple: moisture and cellulose.

Most often this is wood in soil which they hope will remain moist. Dehydration is a major impediment to the success of a colony. The adjoining photos show how once-rural land becomes a housing subdivision and small wood fragments of the original trees remain around the building site until they are finally covered over by new turf or garden soil ... a perfect haven for home-seeking termites. See the last paragraph of 'Not all termites are the same' on page 30.

A typical housing development site; all the trees have been bulldozed, the land graded and the lots marked out.

Smashed fragments of trees remain all over (and under) the site. After construction, covered by buildings and turf, they create near-perfect conditions for colonising termites.

This photo shows the distribution of established termite nests in rural land. Theoretically, this spacing may be superimposed and called 'normal' once a housing development is completed.

A developing reproductive (future king or queen).

A nymph, which stays close to the nursery section of the colony structure until it changes into either a worker, soldier or developing reproductive.

An almost fully developed reproductive termite still to produce wings and eyes.

A soldier, which protects the other castes in the colony.

There is no worker caste shown here. Workers are blind and asexual, and not only build and maintain the colony but find, fetch and regurgitate food to feed the castes shown here.

A fully winged termite, which leaves the parent colony to set up its own.

Termite castes

Each caste has a specific function. After selecting the initial nest site, the king and queen mate and tend to the hatching nymphs themselves. The first nymphs become workers and take over the feeding and nursery duties so the king and queen can solely produce eggs. A three-year-old queen in a vigorous nest could be laying more than a thousand eggs a day. In several species, queens can live more than 25 years and increase in length from around 8 mm to around 30 mm due to an enlarged abdomen with its specialised ovarian system. When the queen dies or degenerates, her place may be taken by supplementary queens selected from the developing reproductives.

WORKERS do all the work. Apart from nursery duties, they forage for food, usually from many sources in many directions. They perform a continual cycle of chewing off wood, partially digesting it on the often long way back to the nest, then regurgitating it for the sustenance of all the other inhabitants: nymphs, soldiers and royalty. The almost indigestible cellulose is broken down by protozoa in their gut. If the wood

< A queen termite of Nasutitermes exitiosus. Her length has increased from 7 mm to 25 mm due to ovarian development.

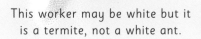

This worker may be white but it is a termite, not a white ant.

A Coptotermes acinaciformis queen. Notice the size difference between the queen and the other termites; however, her head is the same size as theirs.

The difference between the worker and soldier heads is obvious. The soldiers' jaws are for defence only; they have to be fed by the workers.

is moist and partly infected by fungi (quite normal in the moist conditions termites prefer), digestion is further aided. Termites give food from such sources a much higher priority. Anything like that around your place?

SOLDIERS appear next. Once the immediate food and labour requirements are being met, some of the nymphs begin to develop into soldiers, with brownish heads and mandibles specialised for fighting in defence of the colony. In some species, soldiers are also able to emit a white latex-like solution from an orifice called a fontanelle in their 'forehead'. If a working gallery is breached or flight slits are made, the soldiers crowd to the opening, blocking it with their heads until workers can begin resealing with a mud/wood compound. Another defence strategy in some species is to jerk their heads to make a sound somewhat similar to the winding of an old-fashioned watch. If termites are hollowing out, say, an architrave or a window frame and you tap or thump it, you may hear the tiny ratchet sound of termite soldiers snapping their heads backwards and forwards to indicate their defiance.

ALATES or REPRODUCTIVES only begin to be produced once the colony matures. Emerging from the nymphal stages, they differ in appearance to workers by having eyes, wing buds and a slightly longer and flatter abdomen. They are found throughout the galleries until about November/December, when they darken in colour, the wings become fully formed and they assemble at the launching site awaiting optimum conditions for the survival of the species. It's just as well for us that around 99 per cent of them fail.

A Nasutitermes sp. soldier at a breach, defending its territory. The tip of its flask-shaped head can emit a latex-like solution to deter enemies. The mouthparts are small.

Behavioural patterns

About 95 per cent of termite colonies are outdoors rather than inside or under buildings. However it is generally agreed that more than 80 per cent of mainland Australian houses are within 25 metres of a termite colony. The size and age of colonies vary along with the seriousness of the threat. Some may not last long due to lack of food, unreliable moisture or the removal of firewood, garden stakes and so on by humans.

A radius of activity extending 50 metres is not unusual and many houses within that area can be attacked by the same colony. Underground tunnels are constructed to reach the food zones. These tunnels are usually in the top few centimetres of the soil, hardly ever any deeper than 200 mm unless they are following a root or eating out a post. Some species will construct tunnels on top of the ground, particularly in hard, dry and stony conditions. These tunnels may not be populated in the midday heat. The tunnels are also called 'leads' by the pest management industry.

If termites are accidentally or purposely discovered and their active galleries exposed, the disturbance immediately triggers defensive and survival reactions. If the created opening is small enough to be repaired, the soldiers appear and try to guard it from their traditional enemy, the ant, until the workers have repaired it. If ants attack in sufficient numbers and the termites seem to be losing the battle, a rearguard begins to block off the battle from the nest. This may sacrifice thousands of termites that can no longer get back, but the colony survives and the

Like miners, termites leave some 'slices' of timber intact while eating out the rest. Mud packing is also used to prevent collapse.

Termite pest species

queen may increase egg production to compensate. The same strategy is put in place if you discover termites and get so carried away with horror and curiosity that you rip open skirting boards, door frames and maybe even wall linings. Such devastation will result in the death by desiccation of thousands of termites; but don't feel good about it: the nest will survive, the dead will be replaced and they still know where you live! There may be alternative routes; in fact they may already be invading along several other routes.

None of the termite pest species have common names. Those in the pest industry often refer to various groups by an abbreviation of their genus names, such as Copto's, Schedo's, Nasutes and Masto's for *Coptotermes*, *Schedorhinotermes*, *Nasutitermes* and *Mastotermes*. Identification relies mostly on soldier characteristics, including head and mandible shape and serrations.

The most destructive species and where they are found in Australia are:
Coptotermes acinaciformis (most states)
Coptotermes frenchi (eastern states)
Coptotermes raffrayi (Western Australia)
Mastotermes darwiniensis (tropical north)
Shedorhinotermes intermedius (most states)
Nasutitermes exitiosis (eastern states).

A developing reproductive (top), a worker showing a dark abdomen because of the food inside (right) and a soldier (left) of Nasutitermes exitiosus.

Soldiers of Coptotermes acinaciformis.

Coptotermes acinaciformis

In terms of occurrence and total damage caused, this is Australia's most destructive termite species. They often nest in hollow trees and stumps but can set up colonies behind retaining walls, under firewood heaps, in stacked timber, fence posts, and so on. Soldiers are 5–6 mm long with a pear-shaped honey brown head and a fontanelle producing a white bead of defensive latex when disturbed.

Coptotermes frenchi

The soldiers are very similar but slightly smaller at 4–5 mm than *C. acinaciformis*. Their range extends from Queensland through New South Wales and Victoria into South Australia, nesting mostly in trees and using underground tunnels to food sources.

Coptotermes raffrayi

This species is restricted to the south of Western Australia. They make small mounds, often next to trees and stumps, from which they attack buildings. The soldiers are similar in size to *C. frenchi*.

Mastotermes darwiniensis

Masto's are larger insects, with the soldiers up to 12 mm long. Their destructiveness is legendary: of all termites they do the most damage in the shortest time. During World War II, when quartermasters from the south knew nothing about tropical termites, crates of leather army boots were demolished in days, leaving only the metal heel pieces and the eyelets. They have been known to deflate the tyres of vehicles left for a few days in the wrong spot. Leather and rubber are not cellulose but they still attack these products. They will attack any wood in contact with the ground, including shrubs and trees, so orchardists in the tropics have a constant battle. Colonies are usually smaller, with only a few thousand individuals below ground. There are no true workers, this duty being largely performed by nymphs and developing alates. At times groups will break off from the main colony, using a fully developed reproductive as their new queen.

Mastotermes do this sort of damage to palm trees in the tropics.

Schedorhinotermes intermedius

Occurring over most of Australia, the extent and cost of their economic damage is second only to *C. acinaciformis*. There are two sizes of soldier: major and minor. Nests are almost always below ground and are difficult to find. They are very timid and, if disturbed, they usually abandon their workings. They may reopen and return to the gallery weeks later or maybe never. Disturbing this species often means they will just attack from another direction.

Mastotermes darwiniensis is one of the largest Australian species, about 12 mm in length.

Nesting sites

Nesting sites and styles vary within some species; other species are more consistent. Those that usually build mounds, such as *Coptotermes lacteus* or *Nasutitermes exitiosus*, are consistent. So are *N. walkeri*, which nest on the outside of trees. *C. acinaciformis* and *Schedorhinotermes intermedius* almost never build mound nests south of the tropics and the variation of chosen sites is enormous. Text books often say that *C. acinaciformis* 'usually nests in trees or stumps' but this statement is made because those are the first places inspected to find a nest that has been significantly attacking a house. Nests inside trees and stumps grow

minor soldier   major soldier

Schedorhinotermes intermedius castes

large because they are protected. Plenty of *C. acinaciformis* nests are found behind retaining walls and under stacked timber. We tend to forget that many, many nests of this species are never found simply because they began in something as innocuous as a wood sliver, a garden stake or a timber off-cut and the occupants were fortunate enough to find other food sources on which to survive while living deep below soil level for protection. There are a lot of *C. acinaciformis* attacks in inner city suburbs that haven't seen a tree, a stump or a retaining wall for a hundred years or more.

worker   developing alate

∧ Nasutitermes exitiosus often build their dome-shaped mound beside posts.

∨ Some species, such as Nasutitermes walkeri, make arboreal nests.

∧∨ Magnetic termites of the north build a tall, flat-sided mound which runs north–south so that at the hottest time of day, the narrow ends face into the sun. At the cooler times of day, the wide sides are exposed.

∧ A colony of Coptotermes frenchi in the base of this tree has tunnels radiating from here to nearby houses, fences and sheds.

∨ Colonies are often housed in the protection of basal cavities in trees.

∨ This Coptotermes lacteus mound could contain more than a million termites. The species is found along the east coast of Australia.

Not all termites are the same

Australia has over 300 native species of termites, most of which are not pests of timber in service. They are however, important contributors to our environment, reducing fallen trees and dead grass to minerals and organic matter, and themselves being eaten as part of the food chain.

Drywood termites are found in the humid climate of the tropical north and some of these species can aggregate in many small independent groups within structures to communally eat out timbers. Compared to subterranean termites, the drywood termites are usually darker brown and have a thicker cuticle (outer shell or skin) to reduce moisture loss; because of the tropical humidity, they don't need contact with the soil for moisture. Their presence is often detected by finding accumulated dried faecal pellets on horizontal surfaces below the activity. Drywood termites are not a major timber pest, although West Indian drywood termites have been found on many occasions, resulting in million-dollar fumigation jobs including parliament house in Brisbane.

The major economic threats to our structures are the subterranean termites described throughout this book. They need constant access to moisture, which is most readily obtained from the soil. Occasionally they are able to source water from air conditioning towers or from faulty roof drainage that causes continual dampness in the upper levels of buildings. These aberrations are often discussed by pest management technicians when they get together, simply because they are unusual.

Mounds are built by many species. The Magnetic termites up north are grass-eaters only; other mound builders will eat grass and/or solid wood. Some build a nest on the outside of a tree and feed on fallen branches and if it suits them, your house. Mounds are very noticeable and not what every homeowner wants as a backyard feature, but they are easily destroyed simply by mechanically opening them up and inviting ants, birds, lizards and, if you live in the right area, echidnas to finish the job for you.

A flight slit in a tree.

The most insidiously damaging termites are those that begin their nest below soil level; as it gets older and bigger it just goes deeper into the soil, unnoticed by humans. From this nest, they forage in many directions and continue to bring digested wood back to the nest. Because they shun the light and keep their working galleries secure from natural enemies like ants, they eat away undetected until either a timber failure occurs, such as a door sagging because the hinge screws have nothing to hold into, or we notice something amiss and make a closer inspection. More on this in chapter 3.

A *Coptotermes acinaciformis* nest occupied most of the lower part of this hollow tree trunk. Physically removing the nest will obviously eliminate the termites as a threat, but what about the safety of the home and its occupants from the weakened tree?

For the scientific minded ...

Termites are insects in the Order Isoptera. Iso means equal; ptera means wings. Of the insects you may see fluttering around on humid summer evenings, the reproductive termites are the ones with translucent equal-sized fore and hind wings.

Scientific names are of Latin origin and are shown in italics or underlined. The generic name (the genus) is first and begins with a capital letter. The species name is next and has no capital. After the first mention of a name in a text, the genus is usually abbreviated to its first letter followed by a full stop. For example, *Coptotermes lacteus* becomes *C. lacteus* subsequently.

PROTECTING your house

Termites just want to recycle your house. In timber-framed houses, all the construction timbers join up, so if termites can get into a bottom plate or a door frame, these pieces can lead them through to all the wall studs and right up to the roofing timbers.

Nests are almost always outside the confines of a building. This means the design and construction of a house, or the physical and chemical barriers, must force any termites to build their mud tunnels over a surface which can be inspected by homeowners and pest management technicians, exposing the site of invasion.

There are Australian Standards to cover the design and construction of buildings, as well as physical and chemical barriers (see box, p. 44). Council building regulations require construction to be carried out in accordance with these standards as well as any local government requirements.

Design

It is possible to build structures to which termites are not attracted. Steel-framed homes are similar in cost to timber-framed ones; but if you are concerned about the environmental impact, you should know that making steel uses about 20 times more energy than harvesting and preparing timber. Timber framing has recently taken a step forward. Bluepine framing timbers from plantation forests, impregnated with Permethrin, are becoming popular with architects and people building their own homes. The cost premium is only about 8 per cent and this should reduce with more widespread use. Even the paper coating on plasterboard is now offered in impregnated form. Full brick or full block wall homes use less timber, but many still use timber roofing and architraves. If you are building now, design options give you the chance to minimise and maybe even eliminate future termite attack.

Power and telephone entry points are yet to go through the block or brick wall as it is built around this timber-framed house.

The metal caps or shields on piers (or stumps) are designed to make termites show their access tunnels (called leads) if they want to get to the timbers above. In this instance they went up the outside of the pier. If the pier was made of irregular rocks, stone or a timber pole, termites may be able to get up the inside instead, but they would still have to come to the outside in order to get over the edge of the shield.

Bluepine timber frames are partially impregnated with Permethrin insecticide, which remains stable and does not come into contact with the occupants, being inside the walls and roof cavity. It is reportedly not hazardous to mammals.

This construction takes a 'belt and braces' approach, using both a Kordon insecticidal membrane and a steel frame. Steel-framed homes sometimes contain wooden carpet strips, architraves and furniture. The insecticidal membrane can protect this wood.

Steel foundations (with antcaps) suspend the frame for the flooring, which may be a composite sheeting or timber. A suspended-floor design allows access to check for termites and also to maintain or modify utilities (water, drainage, gas, electricity, phones).

A company that provides professional indemnity insurance to the pest management industry revealed that 99 per cent of problem jobs they've had to investigate are from concrete slab on ground construction and only 1 per cent from buildings with suspended wooden flooring. Essentially this is because the underfloor crawl space can be inspected and the termites have to build their mud tunnels up the foundation walls and over ant-capping where they can be seen.

Slabs do crack; just have a look at your garage floor or any other large slab of concrete. Your slab may already be cracked, but this is usually hidden from sight under floor coverings. However, termites can't get through cracks smaller than 2–3 mm, and despite the urban myths, they don't eat their way through. Concrete slabs must be constructed in accordance with Australian Standards before they can be regarded as impenetrable to termites. This is clearly stated in AS3660.1.

The relatively new practice of using styrene waffle pods in concrete slab construction reduces the amount of concrete used, and builders prefer it because styrene costs less than concrete. But because there is less concrete and because termites seem to

Building trends and termite damage

In 1958, Phil Hadlington involved forestry staff and the pest control industry in a survey that indicated 3 per cent of homes had a history of termite attack. This was about the time concrete slab foundations became popular. By 1983, three decades after the advent of slab-on-ground building practices, the figure had risen to 20 per cent. The CSIRO survey of 2004 reported a figure of 33 per cent. Since slabs became the normal construction method, Australian Standards have been developed and the physical and chemical barriers have improved immensely – at least on paper. The recent innovation of concrete-saving waffle pods is likely to make the situation worse again. Irresponsible builders and a few rogue pest control companies have had a bearing on this trend.

The major cause of the termite increase is probably the difficulty in preventing termites getting access to homes built on slabs, either by non-compliance to the Standards during the construction phase or simply because the slab and membrane barrier mean the soil underneath is usually moist, which encourages termites to search for food. Good ventilation under a suspended floor means the soil dries out; there is less attraction for termites, and if they do attack, they have to expose their mud tubes, which are easily discovered during an inspection.

Homeowners who do not renew the chemical barrier at the appropriate time because they don't want to drill and reflood insecticide under the slab and surrounding paths are leaving their homes vulnerable.

The new external monitoring and treatment devices, as they gain popularity, will begin to reduce the number of well-established nests that already surround our homes.

A good thing about having a suspended floor construction rather than a concrete slab is that termites can usually be detected in the crawl space during an annual inspection.

If you really want to beat termites, the indicators are: build or buy a house with a suspended floor that is easy to inspect thoroughly and easy to retreat chemically. Have it inspected at least annually. And put some monitors around it.

like styrene, cracking is almost always a concern. Recently a home mechanic jacked up his car to do some work underneath it. The base of the jack bore the concentrated weight of the vehicle and it broke though a thin part of the concrete slab; he was pinned for some hours until his wife came home.

Because of the time and cost savings, builders will use the waffle pod method unless you specify something different. We suggest you look beyond the initial cost savings; you'll just pay the termite-associated costs later.

It might be against the trend, but if you're serious about avoiding or minimising termite attack, build a house with a suspended floor and use Bluepine treated framing timbers; you'll find it less costly over the life of the house. It probably won't become one of the third of homes attacked (see box, p. 35), so that will save you at least $10 000 in treatment and repair costs.

This physical barrier (an antcap) has done exactly what it was designed and installed to do: cause the termites to build mud protection out from the stump foundation so that their presence will be noted during an annual inspection. If not treated, they could add more mud and build an access lead over the edge of the metal cap and into the flooring joist above. From there, they would be able to eat everything!

Physical barriers

Ant caps on top of piers or stumps were the first physical barriers. Not that termites couldn't build a mud tunnel over them; but they couldn't go through them. Continuous metal strips can also be set into the foundation walls, protruding into the crawl space, to serve the same purpose.

Where utilities such as cables and water pipes go through a concrete slab, there are spaces quite big enough for a termite highway. To prevent this, metal flanges, stainless steel mesh and granite particles of specific size and density must be used. The specifications for physical barriers are found in AS3660.1.

Chemical barriers

There are now two main types of soil impregnation chemicals, with subtle differences. The original is a deterrent chemical barrier that termites will avoid. The second has no repellent effect and termites will tunnel through it, but as they do so, they pick up traces of the chemical. The effect is slow but the colony will die out over a couple of months.

Up until the mid 1980s, the chemical barriers were stable and lasted an estimated 30–40 years. The replacement chemicals last less than 10 years, and manufacturers and pest controllers will only give warranties of 3–5 years for these barriers ... assuming at least annual inspections. (You can't blame them for requiring the inspections, for two reasons: barriers can be breached by human and pet intervention; and, in the event of a claim, the cost of termite damage added to loss of resale value can be extremely high. It's better to find out sooner than later.)

Australian Standard 3660.1 requires the foundation area to be treated prior to a suspended floor being laid; or in the case of a slab, prior to the membrane being laid. It also requires a perimeter barrier to be applied before paths and driveways are put down (after the building is completed). One of the biggest problems has been the oversight (or reluctance) of builders to call back the pest manager to apply the perimeter part of the barrier before the paths go down. If you are an owner-builder, or even the owner keeping an eye on the building in progress, make very sure this perimeter barrier is applied, because it is an essential part of the

Instead of cutting a brick to fit the gap, the bricklayer has expanded the spacing between these bricks. This leaves probable access for termites, which will be below soil level and unseen by the building contractor (and the homeowner).

It's no wonder the CSIRO survey reported that a third of Australia's homes have a history of termite attack. The irresponsibility of some building workers leaving timber offcuts near foundations and the popularity of slab-on-ground construction indicates the percentage could climb.

To avoid damage to electrical and water services, cut and remove the concrete, treat the soil and replace the concrete with pavers, so they are easy to remove for the renewal treatment in 3–5 years.

When a path or patio is drilled, the objective is to force liquid down each hole until it bubbles up the adjoining holes, which supposedly indicates the new barrier is continuous.

Standard. If the perimeter treatment is not done, the builder cannot claim to have constructed in accordance with the Australian Standards. The above photo shows an example of shoddy, almost criminal practice by the building trades. Remember, it is your house; try to visit every day during construction if you can.

If a house is attacked by termites after construction is completed, a pest controller will want to attempt to renew the barrier by drilling holes in the concrete floor throughout the house and the surrounding driveway and paths. Naturally, this labour-intensive operation is costly and very inconvenient to the occupants. The drilling is followed by high-pressure application of chemical into the closely spaced holes so that hopefully the pools of liquid beneath the slab join up or intermingle to renew the barrier and cut off the access the attacking termites were using. This join-up cannot be guaranteed, because you can't actually see what happens below the concrete. Around the house, it is much better if the concrete is cut and lifted; after treatment, pavers can be used to replace it. This of course also makes

A small hand puffer can effectively distribute chemical dust within timbers excavated by termites.

A Termguard reticulation system is installed just before adding the membrane, steel reinforcing and concrete, but after the utilities are in place.

The renew point for a Termguard system is placed outside the house before the paths and landscaping are completed. See the piping against the wall to first apply and later renew the perimeter chemical barrier.

it much less expensive to retreat in another 3–5 years (see photos).

Remember back in chapter 1 where the frenzied disturbance of newly discovered termite activity was described? If they have not been disturbed and there are sufficient numbers of termites to treat with a slow action chemical contaminant introduced into the galleries, the internal drilling, pressure treatment and inconvenience may not be as necessary.

Because the current barrier chemicals do not last more than a few years, reticulation systems which can be installed before the concrete slab is poured have been developed. These systems should also be installed underneath any outside driveways and footpaths. They allow the barrier to be 'topped up' in the coming years. It may be an added building expense, but there is no other convenient way to replenish a barrier under concrete.

An Australian innovation is the development of Kordon, a chemical 'security blanket'; this impregnated webbing can only be applied by accredited technicians during the course of construction. It goes

Monitors and treatment devices

under the whole slab and foundation areas and doubles as an approved moisture membrane. Its big advantages are its longevity and the fact that it is installed by the people who offer the warranty. In 2005 the data for Kordon protection had progressed to 17 years and indicated an extrapolated useful residue for more than 50 years. An even more recent development along the same lines is the Homeguard barrier membrane, which uses Bifenthrin.

This insecticidal Kordon membrane is a barrier to moisture and termite attack. It is laid by professionals once the site is prepared and the utilities are in place. The next step is to lay the steel reinforcing and then the concrete slab (two stages in this split-level home).

As you can see from the previous section on chemical barriers, you cannot be confident of protection beyond 5–8 years from each application or renewal. Fortunately, the fortress approach is now being supplemented by monitoring and aggregation devices that can find and kill off colonies outside buildings. The purpose of aggregation is to attract large numbers of termites to a device and treat them with minimum disturbance so they will continue to perform all their usual activities, such as harvesting food and grooming workmates, thus carrying the treatment contaminant back to the nest where it will affect and kill nymphs, soldiers, developing reproductives and royalty. In other words, the treatment uses the termites' natural instincts against themselves.

If you just want to know whether termites are marauding through your backyard, a simple scattering of timber off-cuts through your garden will provide you with an answer within 2–3 months, often less. However, turning over a block

Termites have found this homemade peg monitor; however, there are too few termites to be confident it would be a successful treatment station.

And it's not just the first thousand or so termites that receive a dose of toxin; it is much better if the rotating 'shift workers' that come to the device over the next few weeks continue to harvest contaminated food and take it back. This is particularly important if the timid *Schedorhinotermes* species is involved.

There are various types and brands of monitoring and aggregation devices available. One is a short dowel rod of Tasmanian oak on a cord, which is inserted into the interior walls. By unscrewing the cover plate and pulling on the cord to view the timber dowel, homeowners can see whether termites have arrived inside the walls. This type of monitoring tries to discover termites after they are already inside the house and hopefully before serious damage occurs. If a dowel shows termite damage, a thorough follow-up inspection and treatment is required from a licensed technician.

Major monitoring systems such as Sentricon and ExTerra can be installed, monitored and used to apply treatment only by licensed technicians. These monitors are set into the ground around the outside of a building. The cover plate is at ground level and inspection is performed by opening it with a special key; if termites are present,

may only reveal 20–50 termites. That's too few to rely on to carry a dose of toxin back to headquarters. And no matter how carefully you turn over any block, you will have broken the connecting access tunnel back to the nest. Wood blocks and anything else that cannot be checked without disturbance are not a treatment option.

Unless you can treat massed termites without disturbing them, aggregation is of little value. The really successful devices are big enough to hold thousands of termites and designed so that treatment can be applied without frightening them off.

Do-it-mostly-yourself monitors are available directly to homeowners. They are loaded with cardboard, which termites find easy to harvest, so they turn up in large numbers making successful treatment easy.

Termites enter through the slots below ground level. Most termite action is in the top few inches of the soil.

The window at the top of the TermiteTrap exposes the cardboard to the light, which termites hate. Because they also want to eat that cardboard, they instinctively pack mud in the gaps to block out the light. This mud indicates termite presence to the homeowner. A phone call to the local pest management technician results in an almost foolproof elimination of the colony at their minimum callout fee.

a contaminant is added by the technician and the lid replaced. Disturbance is minimal and the treatment usually succeeds.

Do-it-yourself monitors have been used by homeowners for years. Ranging from blocks of wood to holey ice cream containers filled with paper, cardboard or wood chips, these monitors can be set into the soil at many points around the home. Any foraging termites from nearby nests will probably find the monitors, but the possibility of a successful treatment is low: first because the monitor has to be opened and treated without disturbing the feeding termites, and being below ground level this is nearly impossible; second, because thousands of termites need to be contaminated, not just a few hundred.

A recent Australian innovation has become commercially available direct to homeowners. The TermiteTrap is a hollow moulded plastic post set 250 mm into the ground but standing about 600mm above the garden bed. It looks like a garden light with a window at the top. The patented window causes the termites to follow their instincts and begin to block out the light using mud. This mud is easily visible through the window. The window also allows inspection without disturbing the termites at all. Holding 6 litres or thousands of termites, this monitor becomes not only an effective do-it-yourself monitor but an efficient treatment device.

All monitors need to be regularly inspected and when termites are detected, a timber pest accredited pest management technician must be called, because the slow action chemicals required to successfully kill off a colony are not available directly to homeowners. There are plenty of insecticides that will kill termites, but the strategy is to have workers carry a slow action contaminant back to the colony so that the queen and nymphs die. This will not happen if termites die before they get back.

Termites may have an inclination for classical music. They attacked this violin and the accompanying carton of sheet music.

WHAT HOMEOWNERS CAN DO TO PROTECT THEIR HOUSE

There is a high probability of termite nests being around your home (see box, p. 44), and termites scout in many directions in search of alternate food sources. If they can gain access, your house joins the menu.

Subterranean termites prefer cellulose in contact with moist soil. If you have gardens or shrubs along your walls, the soil will be moist. You should ensure the garden beds and any mulch are well below the level of the ventilating weep holes and the damp course, and consider installing monitors.

Do not use untreated timber for garden edges or retaining walls, but if these are already in place, put monitoring and treatment devices close by.

If you must store timber (thinking it might come in handy one day) put it up on racks, allowing room to easily inspect underneath. The same applies if you are storing anything else that contains cellulose, such as books, paintings or old tax papers.

Do not affix a lattice, pergola or cubbyhouse to the house in a way which could provide a 'bridge' across chemical barriers, from untreated soil to the house.

Reducing invaders to your 'fortress' home

Since the first use of antcaps and especially since the 1950s when Dieldrin, Aldrin, Chlordane and Heptachlor offered an effective chemical barrier that would last for decades, the pest management and building industries have promoted a fortress approach to homeowners. We tried to build 'termite proof' homes, incorporating various physical barriers and cradled on a continuous envelope of chemically treated soil. Countless hours were spent at Australian Standards committee meetings devising ways and means to prevent termites gaining access to buildings. Yet our fortresses keep failing.

An educated guess is that around 80 per cent of Australian homes are within 25 metres of a termite colony. The colony may only be a few months old, but if it survives it can become a major threat to homes that depend on physical and deteriorating chemical barriers.

Since the late 1990s the pest industry has widened its scope from a fortress-only approach, and as a consequence has gained increased credibility. Combination monitoring and treatment devices installed around homes and buildings are directed at eliminating these surrounding colonies before they find access to the building. You cannot assume that these devices will kill off *all* nearby colonies, allowing you to completely dispense with barriers and inspections, but monitors *are* discovered by termites foraging for new food sources and the treatment is usually successful.

The cost of a termite attack to your home is not measured by adding together the cost of the treatment with the cost of repairs; you also have to factor in the loss of value when you want to sell. Not many people want to buy a house with a history of termite damage, or if they do, they want to pay a whole lot less.

You'll probably save in the long run by continuing to renew chemical barriers if practicable, having at least an annual timber pest inspection, and installing monitors as well.

> ## The Australian Standards
>
> There are three Australian Standards mentioned in various sections of this book:
>
> **AS3660.1** focuses on the physical and chemical barriers used during the construction of buildings.
>
> **AS3660.2** focuses on treatment of termites in and around buildings and includes inspection procedures.
>
> **AS4349.3** focuses on pre-purchase timber pest inspection procedures and reports and therefore includes borers and decay. This Standard requires the inspector to have at least the specified level of training and experience and also requires the report to be presented in a standard format. Any work that is recommended is to be done in accordance with AS3660.2.

INSPECTING for TERMITES

CHAPTER 3

Inspecting for signs of termites is not only good sense, it is essential. If you own a home you should do some regular inspections yourself and also have a professional inspection at least annually. If you are about to sell your home, consider having it professionally inspected even before you talk to your real estate agent. This will ensure there are no ugly surprises just as contracts are about to be exchanged. And if you are buying a home yourself, it is absolutely imperative to get a pre-purchase timber pest inspection. If you need to sign the contract before this can be done, ensure there's a clause making it dependent on the results of a professional timber pest inspection.

What homeowners can do

Without moisture meters, thermal imaging infra-red cameras and other devices available to pest managers, you cannot expect to be as thorough. However, having read the preceding pages and looked at the photos, you will understand that termites almost always travel in from outside a building and that once they find their way inside, all the timbers link up so they can get to almost everywhere in the structure.

Termites hate light, so they will put mud in the cracks of any timber they are eating and will not breach the surface even though they may have hollowed it out to the extent that only the paint is intact. If they have to cross masonry, or timber they don't find edible, they build their typical mud tube over it to get to the preferred food. Your inspection will be based on this information.

Termites have come from under the house, up the studs and are heading for the roofing timbers.

Large earthen masses may be built by termites under houses, or in wall and roof cavities. They provide a half-way house between the nest and the feeding areas and allow termites to rehydrate and regain warmth.

You will be looking for mud tubes and for mud in cracks. You should also tap or run the handle of a screwdriver over the surface of all accessible timbers, listening for any hollow-sounding areas.

You should start with a general overview of your home from the outside. Assess the likely sources of termite nests, such as trees, retaining walls and moisture sources. These potential sources might guide you in deciding where to concentrate your inspection on the inside areas, once you get to that stage. Then check the house itself and any fixed adjoining structures. You are looking for any timbers that are even slightly bowed, buckled or wavy and, as mentioned above, for any mud packing in joints or cracks. Weatherboards can often look warped, and a closer inspection and tapping for hollowness is required.

This lead from a very mature and vigorous nest has engulfed a piece of timber initially used to support the pour base for the concrete floor of the bathroom above. The timber should have been removed. Moisture from these old-style drainage pipes would also have contributed to the invasion.

Now check the outbuildings, fences, garden stakes, pool house, timber garden edges, retaining walls, any stumps, dead trees or shrubs and the firewood heap.

If you have a suspended floor and you can access the underfloor area, you will need a bright light. Enter the area below the house and examine the foundation walls, moving continually in the same direction (either to your left or right) until you are back at the access point. You are particularly looking for mud tubes coming up from the ground over the foundation walls. It almost always seems to be the case that the lowest crawl space is likely to be the dampest and therefore the most conducive to termites. Look at all the ant caps and particularly along the timbers adjoining any solid-fill floor areas such as bathrooms. Anything containing cellulose that is stored under the house should be visually checked before it is moved. You don't want to risk disturbing large numbers of treatable active termites.

The inside inspection should begin at the frame of the front door; keep moving continually to your left (or right) looking carefully at and lightly tapping architraves, door and window frames, skirting boards and so on, until you're back to where you started. Be particularly vigilant in rooms that have plumbing. If there are any leaks that may increase moisture under a bathroom, laundry or kitchen, termites are more likely to focus on such areas – and so should you.

If you can access the roof void, your inspection technique will change; it is impractical for you to reach and tap every one of the hundreds of battens, rafters and ceiling joists. However you still work around the roof in one direction using the brightest torch

This stacked firewood won't fall over. Termites have sealed up most of the gaps to keep out light and ants.

Emerging from the crack between the path and the wall, these termites sought added protection behind the water pipe.

A screwdriver handle can be used to check for hollow sounds without damaging the surface, and the blade to pry into cracked or irregular timbers.

What the professionals can do

Not every pest management company has technicians qualified to inspect for timber pests. Make sure you find out before engaging someone to do an inspection. (See Australian Standards box p. 44.)

Professionals have equipment which allows them to provide a better-quality inspection than you can do yourself, but these tools are still a supplement to visual inspection and sounding. Variations in moisture or heat can be caused by factors other than termites, so inspectors will use their knowledge and experience to give extra attention to the areas which are conducive to termites.

Professionals will follow much the same inspection procedures and routines as suggested for you. They will probably use a Donger to sound the timbers; these are longer than your screwdriver and can reach timbers further away. Suspicious areas can be probed with a small diameter tool with a bright light and mirrors that may be inserted into wall cavities allowing the inspector to verify the presence or absence of pests. Listening devices can be used

or light you have to look for mud in cracks and joints or along the outside of the timbers. Again, be particular over the 'wet' rooms, next to any chimneys and around the eaves. It seldom happens, but termites could use water chronically backed up in the outside guttering to obtain sufficient moisture to support a nest.

Before you embark on such a full inspection, you should read the chapters on borers, decay and defibration so you can check for these at the same time.

Coptotermes sp. termites brought in all this mud packing in just one week.

to detect the sound termites make while moving but particularly when they defiantly respond to a thump on the wall or timber. (*Coptotermes* species do this, but not all species react in this way.)

Thermal imaging cameras of varying quality are also used to determine variations in the surface temperature of walls using infra-red technology. These cameras do not 'see' inside the walls. Temperature variation that might be the result of termite activity within a wall stud or cavity is conducted to the surface and will show up in the image. If your air conditioner is on or if the weather is hot or humid, the variation which would otherwise be caused by termite activity is nullified by the ambient room conditions dictating the wall surface temperature. The below-the-surface influences on temperature will not show up in the image.

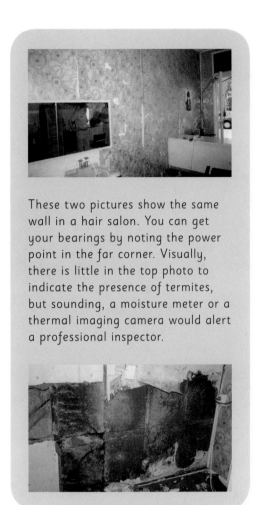

These two pictures show the same wall in a hair salon. You can get your bearings by noting the power point in the far corner. Visually, there is little in the top photo to indicate the presence of termites, but sounding, a moisture meter or a thermal imaging camera would alert a professional inspector.

Alf the wonder dog was trained to detect termite presence.

Photo courtesy of Shane Clarke, Pestforce, Sydney.

Like all tools, the devices discussed above need to be used correctly and interpreted intelligently.

Another inspection 'tool' is a hand-held computer with software especially written for reporting various types of inspections. Following the prompts ensures all relevant procedures and factors are considered during the inspection, and the technician can add comments, recommendations and even explanatory diagrams.

Exclusions

If you have attempted an inspection of your home, you'll know how difficult it is to see everywhere. Even with the best thermal imaging camera in optimum conditions, a professional would be unable to pick up the beginnings of termite activity in a wall stud if the termites were working along the surface nearer the brick veneer than the internal plasterboard.

Some areas, particularly within skillion and cathedral roofs, cannot be accessed. Some subfloor areas can pose problems. False floors and ceilings can prevent a full inspection.

If you have heavy or built-in furniture or cupboards, and your wardrobes are packed so full the inspector cannot get to the wall or floor, these will also be noted as areas excluded from the inspection.

If an inspector finds signs of trouble in an area adjoining a no-access zone, he or she may request that a builder be employed to gain access, so that the questionable area can be properly inspected. If you don't organise this, the onus of risk falls back on you. Of course, in the instance of a pre-purchase inspection, only the vendor can authorise invasive access; if this is withheld, you should talk to your inspector and then decide whether, in light of the risk of not inspecting that particular area, it is worth proceeding with the purchase.

This disused fireplace was closed with a painted plywood panel. When it was removed during a pest inspection, a large termite lead connecting to the roof was revealed.

Choosing a pest management company

A few bad pest managers have given the mainstream industry a reputation it doesn't deserve. A qualified timber pest manager has done a TAFE course or the equivalent and completed a structured training program.

Start by asking someone you know who has been happy with their pest controller. Members of the AEPMA are guaranteed to have access to current technical information and training and carry adequate professional indemnity and public liability insurance (if ever impartial arbitration is needed, and a mistake is revealed, the member must make amends or risk losing their valued membership). You can go to the association website www.aepma.com.au, click on 'Find Company' then 'Practitioners', type in your postcode and click 'Search'. Look for the association logo in advertisements in the Yellow Pages.

Three different professional inspections

THE ANNUAL INSPECTION

This regular inspection may be more often than annual depending on the climate and the risk. If you opt to miss a year, you take the risk and may void any warranty.

Essentially this is an inspection of your home without knowing for sure you have a problem. If you have a new home with a warranty on the treatment applied during construction, or you've had a re-treatment done, the warranty terms usually require you to have and pay for annual inspections. If you miss an inspection, you may void the warranty. These inspections are covered in AS3660.2.

THE PRE-PURCHASE OR 'REPORT FOR BUYER' INSPECTION

If you are selling your home, you may wish to assure prospective buyers that it has no timber pest problems by providing a current report; however, buyers are well within their rights to have another inspection conducted by a company of their own choosing. An analogy is someone providing a pink slip when trying to sell their car, but the potential buyer deciding to get an automobile club to inspect it for more than just whether it is roadworthy before they buy.

Australian Standard AS4349.3 applies specifically to pre-purchase inspections. You'd be wise to specify that you expect yours to be done and reported in accordance with the Standard, which also stipulates the inspector's minimum qualifications. Also ensure that the inspector has professional indemnity insurance, so that if the report is wrong you'll receive due compensation.

If some damaged timbers are found during a pre-purchase inspection, the pest management company cannot even begin to estimate the extent of the damage because the building still belongs to the vendor. They usually recommend seeking the services of a builder to determine the extent.

INSPECTION FOR A QUOTE OR PROPOSAL

This will usually be requested following the discovery of termite activity or damage. We cover this in detail in the next chapter.

Repairs and undiscovered damage

If, after purchasing an older home (a newer one might still be covered under the builder's warranty) you decide to renovate by removing an inside wall or three, it is possible you might find termite damage which was not reported. If you find live termites you may have a case against the inspecting company, providing the termites would not have had enough time to do the damage since the inspection. But if there are no live termites, despite the extent of the damage or how soon after taking possession you discover it, there may be no grounds for compensation. You'd be wise to seek specific advice from the industry association or another pest management company, but there may have been no indications at all visible or available to the inspector at the time of the inspection.

It is quite possible for termites to have obtained access at an earlier time and they would naturally eat into adjoining timbers but not necessarily provide any visible indication in the interior mouldings or the roof. The damage may have been caused decades and many previous owners ago. If once there was a mud tunnel through a weephole, or if a garden bed was once high enough to provide direct access but has now been lowered and all signs removed, it can hardly be the fault of the inspector. Even with all the gadgets and tools available, there would not be any increased moisture, heat variation or sounds if the termites were long gone.

This termite mud packing was in a wall cavity behind plasterboard. It is not a nest, because there is no queen, but many would call it a sub-nest. Others call structures like this bivouacs — they provide places for termites to rehydrate on a long journey between the nest and the working or feeding site.

Explanation of inspection report terminology

Some reports list the company's own explanation for the terms used. If so, their explanation takes precedence over what follows. (They should, however, be similar.)

- **ACTIVITY** This means live insects have been discovered during the course of the inspection.

- **NO ACTIVITY** Means there were no live insects found. It does not mean there is no damaged timber.

- **DAMAGE** Indicates there are some timbers damaged by timber pests (specified). There may or may not be any activity on the day of the inspection. It is almost impossible to estimate how long ago damage occurred. Termite workings inside a piece of timber look little different whether it has been 10 years, 1 year or 1 week since termites were working. Even if only a small amount of damaged timber is found, usually there is a recommendation to have a licensed builder called in to determine the extent of the damage.

- **EXCLUSIONS** Access to some specified areas was not possible. (See the section on exclusions.)

- **RECOMMENDATIONS** If the recommendations in the report are not accepted and followed, it could mean you have no recourse to any compensation.

- **NORMAL WEATHERING** Weatherboards, for example, may be decades old, split in places and have slight fungal deterioration at the joints or corners. If the extent of this 'weathering', in the opinion of the inspector, is of concern, a suggestion (such as to repair leaking gutters) would appear in the Recommendations.

A pre-purchase inspection report is written to give you an understanding of the timber pest situation (termites, borers and decay fungi) at the date of the inspection. You will then decide whether or not to proceed with the purchase. If you are borrowing to fund the purchase, the lenders may also be interested in the timber pest situation. They are major stakeholders and may want to check that the value of the property is not in jeopardy. Also, they would want you to be able to meet loan repayments, and this could be a struggle if you need to spend thousands on treatment and repairs.

Bear in mind that the inspecting company also wants to protect itself and keep its professional indemnity insurance premium down. They could try to do this by making impractical recommendations, but they should not try to get out of their responsibility to you by providing a report on which you cannot make an informed judgment. They are not builders, and if they recommend that a builder be obtained to gain access to a particular area, they should generally give a reason why it is important to do so.

WHAT TO DO if termites are found

The homeowner

There's no point expecting to be calm and unworried if you find termites. The potential inconvenience and financial impact is too well known for that.

Try to remember that termites excavate slowly, so another day or week will make little difference. But you do need to get more information.

Whatever else you do, don't open up the workings any further and don't let anyone, even 'experts', attempt to look for the extent of the damage. Remember, disturbed termites will just close off the damaged workings, sacrificing thousands of their fellows if need be, in order to protect the heart of the colony from access by ants (their instinctive enemy). If workings are badly disturbed, that means you've probably lost the opportunity for a technician to treat those active galleries with a contaminant that could be carried back to kill off the queen and the whole colony.

Termites are often found by accident, such as when a broom breaks through a skirting board or a door sags because there's no wood left to hold in the hinge screws. Whether you see live termites or not, just put things back the best you can. You can try to block up any opening with soggy paper or a rag.

Alternatively, if you discover termite mud during your inspection, you can test whether they are still actively working in the area by carefully prying away some of the mud in a crevice or making a small split or incision in timber that sounds thin or hollow. A small, sharp kitchen or pocket knife is all you need. Do not damage the surface because you'll only have to repair it later. Under no circumstances should you attempt to

If you gently pry open timber you suspect contains termites, you will probably see the soldiers crowd to the opening, barring the way to possible invaders with their heads.

open a mud tube going over brick foundations or masonry; it might fall away and disconnect all the activity within the house from the colony. Bad move.

Once you have a narrow opening about 5–10 mm long, wait a few minutes to see if any termites come to the opening.

A typical termite mud tube provides them with protection between the soil and the timbers of the house.

Usually soldiers will put their heads into the slot, barring the way to those imagined ants until workers can repair the hole with mud. If no termites are seen, there are two possibilities: you have either discovered old damage that no longer is part of a working colony, or the termites are just not working in that area today.

No matter how you find the termites or their damage, your next step is to engage a licensed and qualified timber pest management technician to assess the situation and give you a written proposal explaining your many options.

Notice how close termites excavated to the surface of this framing timber.

SOME SUGGESTIONS ON YOUR STRATEGY IN SELECTING A COMPANY

Before you agree to engage someone:

- Ask them if they have a current licence, what timber pest qualifications they hold, and whether they carry professional indemnity insurance.

- Tell them briefly that you've found live termites (or termite damage which may or may not be active).

- Ask what it might cost to have their inspector make an assessment and prepare a proposal. (Few companies give free written assessments, because it takes time to be thorough and there are substantial litigation consequences for getting something wrong or missing something while rushing to save unpaid-for time.)

Many companies will agree to waive the assessment fee if you accept their proposal, but just as there is no such thing as a free lunch, you can be sure the cost will be covered elsewhere. Be absolutely clear on what you agree upon at this stage.

Before the inspection begins, insist they do not disturb any live termite galleries. You might want a second opinion, and the next inspector will need to verify for him- or herself any undisturbed activity.

If a couple of companies will be competing for your business, you

This ceiling joist has been eaten out and instead of carrying the ceiling load, has itself become a load which may bring the ceiling down.

Roofing timbers can be extensively damaged and repairs can be complicated and expensive.

The purpose of a perimeter barrier of chemically treated soil is to ensure termites cannot gain access to the building without being confronted by the barrier. The liquid soaks in beside and under the edge of the slab or foundation wall, and as soil is backfilled into the trench, more liquid is applied to soak into that as well.

might as well let them know before they start. Do not show or discuss the first company's proposal with the second company, for two reasons: the later company may rely more on the first company's inspection than on making their own full inspection; and second, they may just shave something off the first quote when, if they didn't know the price to beat, they may have given you a much lower price.

All companies have a duty of care to offer industry best practices, which obviously includes an inspection to AS3660.2 or AS4349.3. Because of this, they are obliged to offer all applicable treatment options covered in that Standard. You are under no obligation to accept every treatment option offered, but if you don't, there could be a lessening or total elimination of any warranty. It is completely up to you to weigh up the costs and possible consequences.

Some companies may offer to treat and make repairs. You would then need to have the repair costs separated and itemised, with a finishing time frame, so you can decide whether you'd rather get a builder or handyman to do it later. The extent of the repairs and the inconvenience of getting a builder will influence your decision. You may, of course, decide not to have repairs done at all.

Just because damage may have occurred, say in an interior wall, and some studs and noggings are partly eaten, that is insufficient reason to tear off the wall linings and destroy the house, like you may have seen on TV, unless the structure is significantly weakened.

Some proposals may offer a timber replacement warranty which covers you in case their treatment failed for whatever reason. This warranty will usually be dependent on you keeping up the annual inspections and other set conditions.

Other warranties may be offered, or no warranty at all. However there is still a warranty of some degree under your state's Fair Trading law if what you agreed to and paid for is not completed properly.

As with any business dealings, the fine print is there to be read and should be explained to you if you don't understand it before you accept and sign.

Despite the tone of the previous paragraphs, the pest management industry is full of well-trained people who are in business to provide a first-class service to their customers in order to find and solve pest problems. Integrity and a good reputation are uppermost considerations in their constant aim to satisfy. What they really want are customers who are inclined to recommend them to others.

This soil barrier is being applied with a tool that not only delivers large volumes quickly, but can also be used to push into loose soil. This barrier is being applied between the wall and an outside hot water service which had been set on timber (which termites found and destroyed). Pavers were used to cover this treated soil.

So many entry choices! This construction bungle was discovered when a perimeter retreatment was being done. The use of holey bricks is bad enough but there is also room for a termite highway beside the pipes, which almost certainly go right past some timber framing.

Treatment options

TREATING ACTIVITY

If live termites are present, the Standard requires that an attempt be made to treat them in such a way that the workers will carry a contaminant back to kill off the whole colony. This is done before any other treatment so as not to interfere with or destroy the connection between termites within the building and their outside nest.

Success in treating live termites depends on the number of termites available both at the time of treatment and during the following days or weeks. The more termites dosed, the higher the probability of success. Treating termites in one doorframe, or in an architrave and a skirting board, may not be enough. It is important to recognise that there is no infallible way of determining a colony kill. If, a week later, no termites are present in a timber that was occupied before, it could mean success or it could just mean the termites were disturbed by the treatment and have left the gallery vacant for some weeks, months or forever.

Arsenic trioxide is the oldest of the registered products that act this way and is still used very successfully. It is a poison that works slowly and kills when it is eaten. Very little is used, maybe just a few grams in a whole building. Almost all of this stays attached to the inside walls of the galleries within the timbers and therefore is not a danger

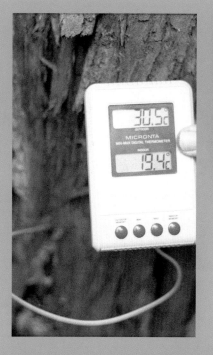

Arsenic trioxide can be introduced through several holes to contaminate the workings.

The probe of a digital thermometer, when inserted into an active nest, will read around 30°C. It is best to take the reading early in the morning, when the ambient temperature is lowest, to avoid confusion.

to people or pets. There is little or no hazard to anyone but the technician (who will be wearing safety gear) at the time of application.

There is a small group of chitin synthesis inhibitors (CSIs) which disrupt normal cuticle formation in the development stages of insects. Chitin is the component of an insect's 'shell' that gives it strength. CSIs may take months to kill off a colony instead of the couple of weeks for arsenic, but the colony is just as dead and will not pose any further threat.

Imidacloprid is a soil barrier chemical which has also been registered as a foam that can be injected into the galleries. The foam bubbles expand along the workings then gradually burst, leaving the termites relatively undisturbed but working in a contaminated environment; they then carry the chemical back to the colony.

BARRIER RETREATMENT

Buildings since the 1960s have generally had chemical barriers applied around the foundations during construction. If termites are alive in your house, they have breached the barrier. The proposal you get will almost always include recommendations for renewing that barrier to once again protect your home from termites. Even if the colony invading your house could not be killed by treating the activity as described above, this barrier renewal treatment aims to prevent further access for a few more years, while at the same time cutting off

The lower area of a tree trunk is drilled with an auger bit; most colonies occur in the basal, root crown area.

Once the presence of an active nest is determined, it is possible to inject liquid insecticide in through the slightly downward-sloping auger hole. Trials are still underway, but eucalyptus oil has been used very successfully to kill nests in an environmentally responsible way. Eucalyptus oils are not all the same, but the authors have found a blend that has never failed.

any termites that may still be inside at the time. Most of these will die of desiccation within a week or so.

Recommendations may include drilling through concrete or tiles in order to reinstall a continuous barrier under and/or around a house built on a concrete slab. In houses with suspended flooring the whole underfloor soil area can be treated along the inside of foundation walls and around the piers or stumps at any time. The foundation walls under filled-floor areas, such as bathrooms, may need to be drilled horizontally to allow chemical application. The same could apply to filled veranda or terrace floors, where drilling from the top may also be required.

The chemicals currently registered for this use may last for 5–8 years but the warranty given is usually less.

Insert a long blade of grass into an auger hole in a tree trunk. If termites are inside, the soldiers will 'attack' the grass and when you withdraw it soldiers may still be attached. The photo shows Coptotermes sp. soldiers.

SURROUNDING AREA TREATMENT

Termites that get into houses are more than 95 per cent of the time from colonies outside the confines of the house.

The inspection should therefore include a thorough check of the gardens, trees, sheds and retaining walls. If suspicious trees are on neighbouring properties or on council land it would be desirable to obtain permission to drill and inspect these trees for termite nests. Even if the termite colony invading your property and threatening your home is outside your boundary, it still makes sense to make any attempt you can to kill it off.

The installation of monitoring devices to intercept foraging termites is a significant step in eliminating colonies, and thereby reducing the threat to your home. These devices cannot be guaranteed to prevent further attack, but if you have enough of them, there is a high probability of success.

TREATMENT SUMMARY

If termites have gained access to your home, you should not ignore the fact. Some treatment is required. If you can't afford or don't want the inconvenience of the full list of treatment options in the proposal, ask what the likely outcome would be if you declined various options.

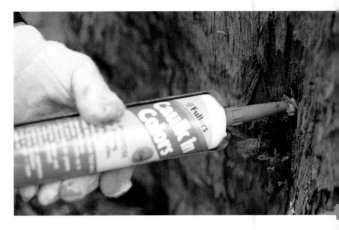

Sealing the holes reduces disturbance and prevents ants getting in to interfere with the treatment.

Comparing proposals

There will always be differences in prices, warranties and recommendations between proposals. Don't forget that you can phone a company back to ask for clarification or explanation. Make your choice based on value, not price – it's your multi-hundred thousand dollar house at risk! A higher-priced proposal may prove to be better value; however, there is no point paying a lot more for essentially the same treatment from someone because you prefer their paperwork or the colour of their uniforms, or for some other insignificant reason.

Masses of termites which had returned to their nest after an arsenic dust treatment.

FAQs on TERMITES

Q: How reliable are moisture meters in detecting termite presence in inaccessible timbers, such as behind plaster sheeting on walls?

A: Termites require high humidity and therefore moisture levels may be higher where they are working than in the adjoining wall areas. Moisture meters can usually detect this variation. But increased moisture levels can also be attributed to leaks, condensation, a faulty damp course, and so on. Inspectors will look for confirming signs whenever moisture is detected. Drywood termites and borers cannot be detected by moisture meters.

Q: How reliable are thermal imaging cameras in the same situation?

A: There are big variations in the available cameras. Some have to be operated closer to the wall than others, meaning that it takes longer to check an area. Those that cannot measure temperature variations of less than, say, one degree Celsius, are not as effective as those that can register a difference of 0.2°C or less. The other limitation is that they work best at an ambient temperature between 19°C and 21°C. As the cameras can only 'see' the surface, air conditioners and heaters must be turned off for hours before the inspection; otherwise the cooling or heating will have more effect on the surface than the conducted heat from any termite activity, which would be well below the surface.

Termites made themselves right at home in this small unoccupied flat at the rear of a home.

Q: I live in a townhouse in a complex. Can termites get access to my home from adjoining townhouses?

A: Most townhouse designs separate units from next door with a solid wall of bricks or concrete blocks. Brickies and block layers don't always completely close the gaps, and this could allow termites access. Also, the roof is usually common, with the timbers above one townhouse continuing above the adjoining townhouse. The body corporate is generally responsible for the common areas and, as most termite colonies that attack buildings come from outside the building, the onus is probably on the body corporate to do their best to minimise termite attack to the strata titleholders' sections. Monitoring and treatment devices set outside in the gardens and the common areas are well suited for this purpose.

Q: Do monitoring devices protect houses?

A: They can alert homeowners to the presence of termites and they provide an ideal way to eliminate a colony which has found its way to the monitor; but of course it is possible for termites to find their way into a house without invading a monitor. All that can be claimed is that by putting monitors in those areas where an invasion is most likely, there is a high probability that termites will find them and give their presence away, allowing effective treatment.

Q: Are there termite species which do not damage timber in buildings?

A: Yes. It is essential that you have the species identified so that appropriate control options can be evaluated. If you find termites in garden mulch, garden stakes, compost or a fence, they may be of a species that will not attack the seasoned, dry timber used to build houses.

Q: Is pre-treated timber resistant to termites?

A: In 2005, arsenic was removed from the pre-treatment formulation for certain 'intimate contact products' such as childrens' playground equipment and tables, but not for framing timbers. These timbers are still termite-resistant. In large pieces of timber, the treatment may not penetrate through to the middle, so if the ends of such pieces are cut off to shorten them, termite access might be possible through the newly exposed, untreated centre. Bluepine and LOSP-treated pine framing timbers are impregnated with the very low-hazard

< The protection, food and moisture provided within a hollow tree ensure the colony survives. From this secure base it can attack many of the surrounding houses.

Electrical failure or fire may occur if termites eat through plastic coating or conduit.

Permethrin. Specifying these frames will not significantly increase your building costs and both termite and borer protection should last well beyond the 20 year warranty offered.

Q: Is soil contact always necessary for termites attacking timber?

A: Most infestations in buildings come from colonies in trees, stumps and landscape materials, but if a reliable source of moisture is available other than the soil, a colony can survive and flourish elsewhere. Colonies have been found in boats and barges, mooring posts in harbours, and in multi-storey buildings where moisture originated from drainage problems in the roof, light wells or air-conditioning towers.

Q: Is arsenic dust treatment better than the more modern chemicals because it kills a colony faster?

A: A colony killed by arsenic trioxide is no 'deader' than one killed more slowly. Even though the termites may live for another few weeks, they probably won't be eating quite as voraciously during this time. The additional damage is almost insignificant – except in the case of a *Mastotermes* infestation in the tropical north.

Q: Are physical barriers such as stainless steel mesh and fine-grade granite stones completely effective?

A: The stainless steel mesh must be of particular size and quality, as stipulated in AS3660.1 (section 6). Lesser-quality mesh has shown unacceptable deterioration, which could allow termites access. The barriers are only as good as the person who installs them and the timing of the installation. Disturbance by plumbers, electricians, telephone technicians and even landscapers after the barrier has been installed could allow termites through.

Q: There are stories of termite colonies lasting for more than 30 years. Would it be the same queen?

Transmission or utility poles which pre-date the use of metal or chemically treated poles are susceptible to attack. They may become hollow and harbour nests that can attack nearby houses.

If a tree in this condition was near your home, you should kill off the nest then entirely remove the tree for safety reasons.

A: Queens can live for many years, maybe 30 or more, but in some species, when a queen is failing, a couple of developing female reproductives may replace her. The new queens have wing buds rather than wing 'stumps' because they never grew full wings and never left the nest.

Q: There is woodland just beyond my boundary fence. How far will termites come from a colony to find a way into my house?

A: There is no Guinness record, but 50 m would not be beyond the scope of a major nest. Termites in wooded areas don't necessarily have to travel far, because they can find so much to eat from fallen branches, but don't take too much reassurance from that because it only takes one colony to find a way into your house for there to be substantial damage.

Q: If most of the termite species that eat the seasoned timber in houses prefer to nest in trees and stumps, why is there so much termite activity in suburban areas where trees have been gone for years?

A: A colonising pair of termites that is 'lucky' enough to find a hollow tree or a stump are likely to be successful; there is plenty of food, a constant moisture source, and protection from human intervention. Colonising couples that get a start in less well-protected sites, such as behind a retaining wall, are still likely to succeed. But those who can only find a garden stake, wood chip mulch or a wood sliver from the developer's bulldozer buried beneath the turf have to quickly find new food sources. The main nest stays below ground level, but if in their foraging efforts they find a house, their food supply becomes assured. They may even find another two or three houses.

As native Australian gum trees age, they usually become hollow. If colonising termites can gain access through fire damage or a hole from a lost branch, they'll find all they could ask for in terms of protection, food and moisture.

Earthen material on the bark of trees, particularly near the base, often betrays a colony.

This interesting cross-section of a tree shows where branches developed. Being denser and harder to harvest, the termites which hollowed out the rest of the decaying pipe have left them intact.

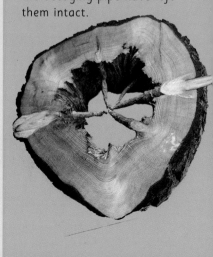

BORERS

CHAPTER 6

The extent and incidence of borer damage in Australia is insignificant compared to termites, but the statistics are irrelevant if borers are eating timbers in your house.

Just because you have some wood with small holes showing, that doesn't mean you have a borer problem; some borers make their holes while the timber is still out in the forest, and others are no longer interested after the timber has dried out. Some borers will never reduce the structural strength of timber ... but there are some types which will completely destroy it. The purpose of this chapter is to explain which borers are which; and what, if anything, you should do about them.

There are three common types of borers, all beetles. Superficially the holes are similar – round, and approximately 1.5 to 2 mm in diameter – but the beetles, their larvae, life cycle, food preferences and their damage potential are quite different.

Pinhole borers

There are many species of Pinhole borers (also known as Ambrosia beetles), and all attack the moist wood of recently felled logs and moist, freshly sawn timber. The female beetles bore deeply (20–60 mm) into the wood to lay their eggs, but do not feed on the wood. The eggs are laid at the bottom of

The fungi-stained holes of the pinhole borer in Australian hardwood.

Powderpost beetles

the bore hole and often the female will die at the entrance to the hole, blocking access to predators of the larvae, which hatch out and feed only on a fungus she also conveyed and deposited at egg-laying time. The fungus thrives in the moist wood and usually causes some staining around the holes.

While there may be several holes, affecting the appearance of the timber, structural weakness is rare and only occurs when there are several holes at one level and the fungus has affected the timber cells at that level.

Moisture is essential to the survival of both the fungus and the insect larvae so that once infested timber dries out, the insects cannot reinfest.

WHAT TO DO ABOUT PINHOLE BORERS

Control in buildings is not needed as the timber is too dry for further damage to occur. The holes made by the beetles may be filled with putty or another filler, particularly in outside areas, to avoid water entering.

Powderpost beetles are pests of structural and joinery timbers in houses and furniture. There are several pests in this group but the main one in Australia is *Lyctus brunneus*.

Infestation is confined to the sapwood, the outer wood of the tree. The heartwood or central area is not attacked. This resistance of the heartwood is due to the absence of starch, an essential food for Powderpost beetle larvae. Only some hardwood timbers have starch content in the sapwood zone, and these are most susceptible to attack, but only during the first few years in service. After this time, all the sapwood other than an outer shell has been destroyed.

Powderpost beetles lay their eggs in the end pores of what are generally classified as hardwoods. Pored timbers would be a better term because not all 'hardwoods' are hard. Pine timbers do not have these pores; instead they have much smaller openings known as tracheids, which are too fine to receive the eggs. Even though they may have starch content, pine timbers are thus immune to Powderpost beetles.

The Powderpost beetle (Lyctus brunneus) is about ▢ mm in length.

A Powderpost beetle larva.

Eggs laid 1–3 mm into the end grain are protected from predators as the hatching larvae begin eating and tunnelling lengthwise through the starchy sapwood. When fully grown, the larvae move close to the surface where they pupate and change into beetles. They emerge through round holes about 1.5 to 2 mm in diameter. The life cycle from egg to adult is completed in 4–12 months. Unlike Pinhole borers, Powderpost beetles will generally reinfest until all sapwood is destroyed.

In Queensland and New South Wales, government legislation makes it illegal to use timber with susceptible sapwood in the manufacture of furniture and joinery or as flooring timbers unless it is treated with an approved preservative. Hardwood framing timbers for buildings are used less frequently today, but if they are, the loss of the sapwood edge must not compromise their structural strength.

When inspecting buildings, you may notice the paler sapwood edges of hardwood frames and other members showing many holes and a fine, flour-like dust spilling out through the holes to accumulate on any horizontal surfaces below. If the building is more than 2–3 years old, no more damage is being done; and if the sapwood proportions have not been exceeded, there should be no significant loss of structural strength.

With the increasing importation of Asian and Indian furniture, including cane pieces, there is the possibility of finding small Powderpost beetles eating the sapwood. Treat with light oil or Permethrin.

WHAT TO DO ABOUT POWDERPOST BEETLES

In most instances, nothing needs to be done to kill the larvae inside the timber because, by the time you discover the damage, there is no more activity and no further loss of strength.

If you believe the manufacturer is at fault in using unprotected sapwood, you should raise the issue of a replacement.

If the sapwood has not been destroyed, you may consider filling the holes with a proprietary filler and then applying paint. If the end-grain pores are unreachable to female Powderpost beetles and the rest of the surface is filled, there should be no more eggs laid; however, a few beetles may still emerge later from any larvae working inside.

This laboratory breeding block for Powderpost beetles shows holes in all but the centre heartwood area, which does not contain starch.

Furniture beetles

There are two main beetles in this category: *Anobium punctatum* is the Australia-wide pest and *Calymmaderus incisus* (the Queensland pine beetle) occurs in northern NSW and Queensland. Most attack is to pine timbers, although some hardwoods are also susceptible. Joinery, furniture and flooring timbers are usually what is affected.

With the extent of the current use of untreated pine for framing timbers these pests could be a much more serious a threat to homes than they actually are. One probable reason they are not is their reluctance to attack dried-out timbers in hot roofs: they are seldom found in drier inland climates. The beetles can fly, but seldom do, so spread is slow even within the same building.

Infested furniture is the usual mode of transfer between buildings, and even from room to room. The presence of their round holes can 'enhance' antique furniture, but this is not desirable if there is still activity, because the whole piece can be severely damaged.

The furniture beetle lays its eggs in cracks and crevices of (preferably)

The Queensland pine beetle

∧ The dust of the Powderpost beetle (left) is much finer than that of the Furniture beetle (right).

∨ The common Furniture beetle is about 5–6 mm in length and has a distinctive peak on the thorax which almost hides the head when viewed from above.

the sapwood of timber. They can also lay eggs back into the flight holes in timber from which they have just emerged. The larvae 'honeycomb' the interior and return close to the surface to pupate. Their emergence or flight holes are round and about 2 mm in diameter. The period from egg to adult is 1–3 years, depending on temperature and humidity.

The most severe infestations are usually found in the

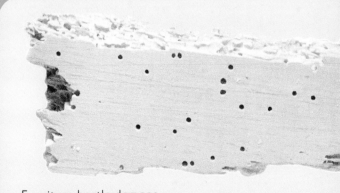

Furniture beetle damage

European House Borer larvae can devour as much in two years in the Australian climate as they can over seven years in Europe.

Doug Howick Gallery

high-humidity areas of the house. Furniture beetles favour pine flooring, particularly where sub-floor ventilation is restricted. Timbers such as Baltic pine, Radiata pine, Hoop pine, Bunya pine and NZ white pine are favoured. Roofing timbers are usually too dry and get too hot. Some hardwoods, such as English oak, are susceptible but the Australian eucalypts are resistant.

WHAT TO DO ABOUT FURNITURE BEETLES

Where flooring is attacked, most of the emergence holes are on the undersurface; therefore an infestation may be several generations old before it is discovered. Treatment of flooring is almost always a job for pest management technicians; they have specialised nozzles and pumping equipment which may be more appropriate if the extent of the attack puts it beyond the scope of most homeowners. But even so, the infestation may take quite a long time to control, because larvae not in contact with the insecticide will have to complete their life cycle and emerge as adults before being affected by the residual Permethrin insecticide on the timber surface. The timber is often replaced, because borer holes in flooring can reduce the saleability of a house.

There are some borer control products available at hardware stores. Essentially, if the surface is not sealed, any oil-based liquid can soak into the already damaged sapwood and kill larvae that are not too deep below the surface. Using a syringe may allow you to get deeper penetration.

Furniture may be fumigated, although this has almost been phased out. Heat treatment can also be effective. Holes should be filled and painted over so that any new holes can be recognised, indicating that the infestation is still in progress, even if diminishing.

European house borer damage in pine timber. The oval emergence holes are 5–7 mm across the long axis.

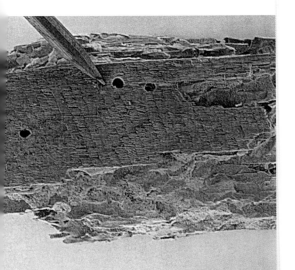

European house borers

These pests were introduced from Europe, arriving in softwood framing timbers. The Australian Quarantine Inspection Service (AQIS) must be notified if this pest is detected. More than 3000 homes and buildings have been fumigated in an effort to eliminate the European house borer. Its European life cycle is around 7 years but Australia's climate would reduce this to 1–3 years. Since it's a very large beetle (20 mm) and therefore has larger, more voracious larvae, allowing it to get a foothold in Australia would spell disaster for many of the pine-framed houses built since the 1960s.

The emergence holes are oval and 5–7 mm across the long axis.

WHAT TO DO ABOUT OVAL BORER HOLES

There are a couple of other, native borers which make oval emergence holes of similar size. If you find such holes, please call your local pest manager who will identify the species and if necessary call in the AQIS.

CHAPTER 7

WOOD DECAY and DEFIBRATION

WOOD DECAY

Wood decay is caused by certain fungi which destroy the cellular structure of timber, often leading to collapse. Moisture is essential, so the term 'dry rot' is misleading.

Timber in service usually dries out to 12–15 per cent moisture in coastal areas and 8–12 per cent in the inland regions. Decay fungi will not develop in wood with a moisture content less than 20 per cent. As builders and architects do not plan for timber to be in high-moisture situations, decay problems occur mostly where ventilation has become restricted or where there are leaks or drainage malfunctions.

Moist conditions are also conducive to termite activity, because decaying timber provides them with wood (carbohydrates), fungal growth (protein) and moisture.

WHAT TO DO ABOUT FUNGAL DECAY

Timber affected by fungi loses weight and strength. In most situations it should be replaced and the source of moisture determined and rectified or eliminated.

Fungal decay occurs in moist situations, such as this poorly ventilated under-floor area. Treatment involves stemming the source of moisture, improving drainage and ventilation, and replacing the weakened timbers.

DEFIBRATION

Wood can deteriorate and collapse due to a chemical breakdown of the cell walls. The most affected timber species are Oregon and Radiata pine, close to the coast where the air is salty. Analysis of affected timber usually reveals higher than usual sodium content and moisture levels.

Defibration can also occur in factories where chemical gases or fumes are present. In houses, vapours from combustion heaters and gas appliances which leak into the roof void through poorly maintained flues have been known to cause defibration much more rapidly than exposure to salty air.

WHAT TO DO ABOUT DEFIBRATION

Replacing the affected timber with a resistant species is the only recommended and effective procedure when advanced defibration occurs. Alternatively and in addition, exposure to fumes could be reduced or eliminated.

Underneath this 'hairiness' was a tile batten affected by defibration, a chemical breakdown of cell walls usually caused by salty seaside air or fumes from gas heaters.

< Decay fungi attack the cell structure of wood. Timber loses its strength and weight.

∧ Oregon tile battens showing the effects of defibration.

This mud is not a nest. There is no queen. It is a bivouac or a rehydration area between the food-gathering area and the nest.

INDEX

AEPMA 9, 50
Aldrin 44
Ambrosia beetles 69
Anobium punctatum 72-73
antcaps 33, 36
arsenic dust 60, 61, 64, 67
Australian Standards 32, 35
    3660.1 34, 36, 37, 44
    3660.2 44, 51, 58
    4349.3 44, 51, 58

bait monitors 35, 40-43
barrier systems 36-40
Bifenthrin 40
Bluepine 32-33, 36, 66

Calymmaderus incisus 72-73
castes 16-21
chemical barriers 37-40, 58, 61
chitin synthesis inhibitors 61
Chlordane 44
Coptotermes spp. 23
    C. acinaciformis 23, 24, 26, 27, 31
    C. frenchi 23, 24, 29
    C. lacteus 24, 26, 29
    C. raffrayi 24
CSIRO 11, 35

defibration 11, 76-77
Dieldrin 44
drilling 35, 38, 62
drywood termites 30

European house borer 74-75
ExTerra 41

foam 60
fungal decay 76
Furniture beetles 72-75

granite particles 36, 67

Heptachlor 44
Homeguard 40

Imidacloprid 61
inspections 45-53
    annual 51
    pre-purchase 51, 53
    proposal or quote 51
Isoptera 31

Kordon 34, 39-40

listening devices 48
Lyctus brunneus 70-72

Mastotermes darwiniensis 25, 67
Microcerotermes spp. 8
moisture meters 45, 65
monitoring systems 35, 40-43, 63, 66

Nasutitermes spp. 21
    N. exitiosus 19, 23, 26, 28
    N. walkeri 26, 28

Oregon 77
Order Isoptera 31
organochlorine insecticides 7, 44

Permethrin 32, 67, 75
physical barriers 36
Pinhole borers 69-70
Powderpost beetles 70-72

Queensland pine beetle 72-73

reticulation systems 39

Sentricon 41
Schedorhinotermes intermedius 23, 25, 26-27
sniffer dogs 49
stainless steel mesh 36, 67
styrene 34, 36
supplementary queens 18

Tasmanian oak 41
TermiteTrap 42-43
thermal imaging cameras 49, 50, 65